Anna Karoeny da Silva Santos
Jackeline Alves Araujo
Marcelo M. Carvalho

Microbiological Profile of Hospital Infections in ICUs

Anna Karoeny da Silva Santos
Jackeline Alves Araujo
Marcelo M. Carvalho

Microbiological Profile of Hospital Infections in ICUs

Imprint

Any brand names and product names mentioned in this book are subject to trademark, brand or patent protection and are trademarks or registered trademarks of their respective holders. The use of brand names, product names, common names, trade names, product descriptions etc. even without a particular marking in this work is in no way to be construed to mean that such names may be regarded as unrestricted in respect of trademark and brand protection legislation and could thus be used by anyone.

Cover image: www.ingimage.com

This book is a translation from the original published under ISBN 978-613-9-73162-6.

Publisher:
Sciencia Scripts
is a trademark of
Dodo Books Indian Ocean Ltd. and OmniScriptum S.R.L publishing group

120 High Road, East Finchley, London, N2 9ED, United Kingdom
Str. Armeneasca 28/1, office 1, Chisinau MD-2012, Republic of Moldova, Europe
Printed at: see last page
ISBN: 978-620-7-88571-8

"The important thing is not to stop questioning."

Albert Einstein

ACKNOWLEDGEMENTS

To God, for **my life, family and friends.**

To the institution for the creative and friendly environment it provides.

To all my teachers for **providing me with** not only rational **knowledge**, but also **the manifestation of** the **character and affection of** education in the process of professional training, **so that they dedicated** themselves to me, not just for **having** taught me, but **for having made me learn.**

To my parents, for their love, encouragement and unconditional support.

My thanks go to my friends from the course, my work mates and brothers in friendship who have been part of my education and who will certainly **continue to be present in** my life.

SUMMARY

INTRODUCTION: Hospital-acquired infections represent a serious public health problem, leading to increased morbidity and mortality among patients admitted to intensive care units, complications related to healthcare, as well as increased length of stay and consequently higher hospital costs. OBJECTIVE: The aim of this study was to investigate the microbiological profile of hospital-acquired infections in intensive care units (ICUs) at a public hospital of reference for high complexity in the state of Piauí. METHODOLOGY: A total of 33 medical records relating to the months of June, July, August and September 2014 with HI were assessed. RESULTS: A death rate of 57.6% was identified, 33.3% related to respiratory system infection, with **Klebsiella spp** responsible for 22.2% of infection cases, which revealed 100% sensitivity to carbapenem antimicrobials (meropenem, ertapenem, imipenem), followed by **Acinetobacter spp** (18.5%), **Pseudomonas aeruginosa** (14.8%) and **Escherichia coli** (11.1%), respectively sensitive to Poliximine B and Tigecycline (100%), Amikacin (100%), and Cefepime (100%), Tigecycline (100%) and Gentamicin (100%). Delayed bladder catheterisation was the most common invasive procedure performed (87.9%) by patients in both ICUs. Vancomycin, Metronidazole and Tazocin were the most commonly used antimicrobials in n=11, 33.3% of cases. CONCLUSION: It can be concluded that the most prevalent topography in both ICUs is the respiratory tract, with the microorganism responsible for the largest number of HI being **Klebsiella spp,** which is mainly sensitive to antimicrobials in the carbapenem class. The following microorganisms were also isolated **Acinetobacter spp, Staphylococcus aureus, Pseudomonas aeruginosa, Enterococcus spp, Enterobacter spp. Escherichia Colí, Proteus mirabillls, Serratia marcenscens, Stenotrophomonas maltophilia and beta haemolytic Streptococcus.**

Keywords: Hospital infection; microbiology; intensive care units.

SUMMARY

CHAPTER 1

INTRODUCTION

Hospital-acquired infections lead to an increase in the morbidity and mortality of individuals admitted to hospital units, as well as an increase in the length of stay and consequently an increase in hospital costs. Currently, intensive care units (ICUs) represent an important and indispensable tool of modern medicine for providing care to critically ill patients using advanced technological resources, but they are also burdened with the risk of hospital-acquired infections (HAIs) (PATRÍCIO, 2008).

The Ministry of Health (MoH) defines hospital-acquired infection (HI) as an infection acquired after the patient has been admitted to a hospital unit and which manifests itself during hospitalisation or after discharge, when it can be related to hospitalisation or hospital procedures (BRASIL, 1998).

In healthcare institutions, inpatients are exposed to a wide variety of pathogenic microorganisms, especially in intensive care units. The ICU is an environment that makes patients more vulnerable to infection risks, both because of their condition and because of the variety of invasive procedures that are carried out on a daily basis (MOURA, 2007).

According to Oliveira et al. (2010), studies have shown that these units are responsible for between 5% and 35% of all HIs, and are also one of the main factors in high mortality rates, ranging from 9% to 38%, with around 60% directly related to the presence of HIs.

In these units, infections are initially associated with the clinical severity of patients, therapeutic diagnoses and interventions, such as the use of invasive procedures (bladder catheterisation, venous catheters, intubations, tracheostomy, mechanical ventilation), patients using immunosuppressants, with chronic illnesses and trauma, with colonisation by resistant microorganisms, and prolonged length of stay in the institution are considered relevant risk factors that can be directly associated with hospital infection (NOGUEIRA, et al;, 2009).

Parallel to this situation, the importance of the manifestation of resistant microorganisms (RM) and infections related to health care stands out.

There has been a progressive emergence of multi-resistant bacteria in the hospital environment over the last few decades, constituting a threat to public health worldwide and related to the indiscriminate and inappropriate use of antimicrobials, whether in hospital or in the community. The situation is aggravating as it contributes to an increase in patient morbidity and mortality and the costs related to their hospitalisation and, above all, due to the reduction in the technological arsenal or the lack of therapeutic options for the treatment of some micro-organisms that cause infection (OLIVEIRA, et al; 2012b).

There is currently no standard definition for multidrug resistance in bacteria, but the term is commonly used to describe resistance to two or more unrelated classes of antibiotics, to which the bacterium is normally considered sensitive. The ability of different species to resist the inhibitory action of antimicrobial agents has become a global problem of increasing prevalence. Resistance to antimicrobials is considered to be a genetic phenomenon, related to the existence of genes contained in microorganisms that encode different biochemical mechanisms that prevent the action of drugs (COUTO, et al; 2010).

According to Couto et al. (2010), normally susceptible bacterial populations can become resistant to antimicrobial agents through mutations and selection, or by acquiring new resistance genetic material from another bacterium. A single change can result in resistance to several different antimicrobial agents or even multiple classes of unrelated drugs.

1.2 Delimitation of the topic

The object of this study is the microbiological profile of hospital infections in two intensive care units. It is therefore believed that this subject should be approached in a reflective and critical manner so that it can contribute to improving the quality of care.

1.1 Objectives

1.2.1 General Objective

id To identify the microbiological profile of hospital-acquired infections in the intensive care units of a high-complexity referral hospital in the state of Piauí.

1.2.2 Specific Objective

V Verify the topography of hospital-acquired infections in ICU patients with HI;

v' Identify the main microorganisms that cause hospital-acquired infections in ICUs;

D Describe the antimicrobials used in patients with hospital-acquired infections in ICUs;

D To describe the sensitivity profile of antimicrobials used by patients with HI in ICUs;

1.3 Justification and Relevance

In Brazil, the magnitude of the problem of health-related infections is not fully known, but a study carried out in 99 hospitals (1995), which is still the most representative nationally, revealed a prevalence of hospital-acquired infections of 15.5 per cent, which is more than twice that established by the Ministry of Health, which is 6 per cent (PRADE, et al; 1995).

According to Cavaleiro (2011), the need for HI prevention and control programmes is reinforced as a means of guaranteeing the quality of care and offering greater patient safety. Routinely carrying out epidemiological studies is important, as it makes it possible to diagnose the situation by characterising the incidence and prevalence of infections, as well as assessing their evolution.

In controlling hospital-acquired infections, special emphasis should be placed on diagnosis, optimised treatment and measures to prevent infections caused by multidrug-resistant bacteria. These strategies should be integrated into a multi-professional approach, with the adaptation of infrastructures, periodic training and awareness-raising actions for all health professionals and the promotion of local antibiotic control policies (CAVALEIRO, 2011).

Bacteria use various strategies to escape the inhibitory effects of antimicrobial agents and have developed highly efficient mechanisms of dissemination, either by clonal means or through resistance determinants. It is becoming increasingly difficult to overcome microorganisms, which have several billion years of learning to adapt to hostile environments, such as those containing antimicrobial agents. Controlling the emergence of pathogens resistant to antimicrobial agents represents a major challenge for the medical community, public agencies and society (COUTO, et al; 2010).

This study is relevant as it contributes to the knowledge of health professionals, given the emergence of hospital-acquired infections, especially in these units related to the emergence of microorganisms that cause serious infections. This way, the study of these microorganisms by the nursing team can be learnt about and encouraged in order to establish goals and discussions on the subject, with the aim of reducing cases of infection in intensive care units.

CHAPTER 2

LITERATURE REVIEW

1.2 Hospital infection

Intensive Care Units (ICUs) are units designed to care for clinically serious patients who need continuous monitoring and support of their vital functions. It is considered a critical area, both because of the haemodynamic instability of the patients admitted and because of the high risk of developing Healthcare-Related Infections (HAIs).

HAIs are defined as any infection that affects an individual, whether in hospital institutions, outpatient care in day hospitals or at home, and which may be associated with a care procedure, whether therapeutic or diagnostic (OLIVEIRA; et al, 2012a).

New microorganisms have been documented and infections have resurfaced with new force, especially in intensive care centres. Hospital-acquired infections are considered more serious in these units, as they are related to the care of critically ill patients who depend on intensive life support.

IHs in ICUs are initially associated with the clinical severity of patients, the use of invasive procedures such as central venous catheters, mechanical ventilation (MV), the use of immunosuppressants for an extended period, the prescription of spectrum antimicrobials and colonisation by resistant microorganisms, prolonged length of stay, among other factors, which favours the natural selection of microorganisms (OLIVEIRA; et al, 2010).

In the study **"Magnitude of Hospital Infections"** carried out by Prade (1995), 1,129 patients were found with HI out of 8,624 patients admitted on the day who had been hospitalised for at least 24 hours, with an average length of stay of 11.8 days. The rate of patients with HI was 13.0 per cent. The highest rates of HI were found in public hospitals (18.4%) and the lowest in private hospitals (10.0%), knowing that this difference occurs because public hospitals treat highly complex cases while private hospitals are more selective, treating less complex cases. These indices showed that the Southeast had the highest prevalence of HI (16.4%), followed by the Northeast (13.1%), North (11.5%), South (9.0%) and finally the Centre West (7.2%) (PATRÍCO, 2008).

In 2001, a multicentre prevalence study carried out in 22 hospitals in Turkey analysed 56 ICUs, with 236 cases of HI reported. Of the patients assessed, 46.7 per cent had one or more episodes of nosocomial infection. The most frequent sites for HI were: lower respiratory tract pneumonia (28%), bloodstream infection (23.3%) and urinary tract infection (15.7%). In 2000, the prevalence rate of HI in five reference and teaching hospitals located in the capital Teresina was published, giving a result of 27.9%, i.e. 12.4% more than the prevalence recorded at national level, which is 15.5% (MOURA; et al, 2007).

Hospital infection rates in ICUs vary between 18 and 54 per cent, and are around five to ten times higher than in other inpatient units in a hospital. The high mortality rates in these units commonly vary between 9 and 38 per cent, and can reach 60 per cent and are directly related to the occurrence of hospital-acquired infections (OLIVEIRA; et al, 2010).

1.3 Main microorganisms causing HI

Staphylococcus aureus is one of the main micro-organisms causing surgical site and bloodstream infections, and is considered the second biggest micro-organism causing pneumonia, cardiovascular infections and bacteraemia. In addition to **Staphylococcus aureus**, **Staphylococcus epidermidis is** also a major cause of increased HAI rates. As a species that colonises the skin, **S. aureus** is often inoculated during invasive procedures or through transmission by healthcare staff and can cause various HAIs (bacteraemia, endocarditis, osteomyelitis, peritonitis). Its pathogenicity has been associated with resistance to certain antimicrobial agents, thus restricting the spectrum of antimicrobials available for treatment (FREIRE; et al, 2013).

More than 90 per cent of all enterococcal infections are caused by **E. faecalis**, although more recently there has been some degree of change in the distribution of enterococcal species responsible for hospital-acquired infections. **E. faecalis** usually shows high-level resistance to aminoglycosides, but resistance to ampicillin is rarely applied. Enterococci are naturally tolerant to penicillins and resistant to cephalosporins, clindamycin and aminoglycosides (COUTO; et al, 2010).

Klebsiella spp is a gram-negative bacillus found in the gastrointestinal tract of colonised individuals. It is an important pathogen that is frequent in hospitals, mainly associated with urinary infections and hospital-acquired pneumonia (COUTO, 2010; MOURA; et al, 2007). Hospital infections caused by **Klebsiella are a cause for** concern, as these microorganisms promote the production of extended spectrum enzymes capable of generating resistance to potent antimicrobials (ALMEIDA; et al, 2012).

Other studies show that over the last four decades the incidence of **Pseudomonas aeruginosa has** accounted for 10 per cent of all nosocomial infections. It has characteristics associated with its natural resistance, the large number of antibiotics used (multidrug-resistant strains) and antiseptics, which makes it responsible for epidemic and endemic infections, being one of the most frequent in hospital infections (FERRAREZE; et al, 2007).

Its most prevalent clinical manifestations are pneumonia, central vascular catheter infections and urinary tract infections, among others. Almost all hospital equipment and materials, especially those with liquid components, can serve as a reservoir for **Pseudomonas ssp,** which, combined with its multidrug resistance, can facilitate its spread throughout the hospital and to the extra-hospital community during patient transfers or post-discharge (ABEGG; SILVA, 2011).

Acinetobacter baummannii, a gram-negative, non-glucose fermenting coccobacillus (GGNF) found in nature and in hospitals, considered an opportunistic pathogen with low virulence potential, is considered an important microorganism that causes HI associated with various infections, the most common being: bloodstream infections, soft tissue infections, urinary tract infections, intra-abdominal infections, meningitis and endocarditis. It has numerous mechanisms of resistance to antimicrobials and is considered a multi-resistant pathogen that is difficult to treat, especially in vulnerable patients (COUTO; et al, 2010).

Stenotrophomonas maltophilia is an aerobic gram-negative bacillus that causes bacteraemia, urinary tract infection, skin and soft tissue infections and endocarditis. It is considered an uncommon agent in infections in immunocompetent individuals, but it has been highlighted as an emerging nosocomial pathogen with high morbidity and mortality, especially in immunocompromised patients. The spectrum of infections by this microorganism is varied and can also cause intra-abdominal infections, ophthalmological syndromes and sinusitis (RODRIGUES; DI GIOIA; ROSSI, 2011).

1.4 Bacterial Resistance

Healthcare-related infections (HAIs), when associated with multidrug-resistant microorganisms, increase the length of a patient's stay in hospital, costs and mortality. Considering these facts, in 2010, the World Health Organisation (WHO) launched the third global challenge for patient safety based on containing bacterial resistance (OLIVEIRA; et al, 2013).

The acquisition of genetic information determining resistance to antibacterial drugs can occur through a variety of mechanisms, which are widespread in a variety of bacterial genera. Fundamentally, the acquisition of resistance by a sensitive bacterium is always the result of a genetic change, which is expressed biochemically. These bacteria can become resistant to microbial agents through mutation and selection, or by acquiring new resistant genetic material from another bacterium. A single change can result in resistance to several different antimicrobial agents or even multiple unrelated drugs (COUTO; et al, 2010).

1.5 Hospital infection prevention actions

In the fight against infection, the following are recommended: hand washing; hand antisepsis, hiring specialised staff, improving nursing techniques, limiting the use of antibiotics when therapeutic and, above all, prophylactic; motivation for treatment, sterilisation, antisepsis, isolation of infected or colonised patients, indication of microbiological examination, analysis of agents and their spectra, preparation of guidelines, immunisation programmes, personal vigilance and patient care (PATRICIO, 2008).

A study by Oliveira, et al (2013) suggests that greater attention should be paid to training, scientific meetings and meetings aimed at addressing bacterial resistance in training involving all professionals. Proposing approaches such as forums for discussion among professionals in relation to the difficulties perceived and perspectives reported can favour a greater understanding of the problem, as well as subsidising the development of goals that seek to achieve maximum adherence to good practices aimed at safer care.

According to Cavaleiro (2011), it is estimated that 20 to 30% of infections could be avoided with prevention and control programmes, which should therefore be implemented and known by the entire healthcare team. Routinely carrying out epidemiological studies is important because it makes it possible to diagnose the situation, characterising the incidence and prevalence of infections, as well as assessing their evolution. It is on the basis of the diagnosis of the situation that preventive strategies can be drawn up and therapeutic measures adapted to the reality of the services.

These strategies must be included in the health policy of each health institution and then adapted to each service according to its specific characteristics. In controlling hospital-acquired infections, special emphasis should be placed on diagnosis, optimised treatment and measures to prevent infections caused by multi-resistant bacteria. These strategies should be integrated into a multi-professional approach, with the adaptation of infrastructure, periodic training and awareness-raising actions for all health professionals and the promotion of local antibiotic control policies (CAVALEIRO, 2011).

CHAPTER 3

METHODOLOGY

3.1 Type of Study

This is a retrospective, descriptive and exploratory field study, with a quantitative approach, carried out in two intensive care units of a high-complexity reference hospital in the state of Piauí.

Descriptive research aims to provide greater familiarity with the problem under study, with a view to making it more explicit or to building hypotheses (GIL, 2010).

According to Gil (2010), exploratory research provides greater intimacy with the problem, seeking without formalities and relatively accessible, in which the researcher seeks to obtain an understanding of the facts that have an influence on the condition that forms the object of study.

The quantitative approach allows all the information acquired according to the classification to be collated in tables, enabling statistical judgement. The variables under study can be quantified, allowing the use of connections and other statistical methods, making it possible to even assess the margin of failure of the effects obtained (GIL, 2010).

3.2 Field of Study

The research was carried out in two intensive care units of a high-complexity referral hospital in the state of Piauí.

This hospital is located in the city of Teresina, the capital of the state of Piauí. It was chosen because it is a public, general, large teaching hospital and a reference for high complexity in the state of Piauí, treating patients from all over the state as well as patients from the neighbouring states of Maranhão, Tocantins, Pará and Ceará.

The hospital has two intensive care units, called ICU-1 (formerly known as the general ICU) and ICU-2 (formerly known as the emergency department ICU). The ICUs have a total of 15 beds, divided into ICU-1 with 8 (eight) beds and ICU-2 with 7 (seven) beds, and have an average of 73.5 (seventy-three and a half) patients per month. The majority of these units admit patients with neurological problems and those after major surgery, and one bed in each unit is for patients in isolation.

The hospital offers care in various specialities, both outpatient and inpatient, and has 427 (four hundred and twenty-seven) beds, distributed in 04 (services) and 11 specialised clinics, as well as a hospital infection control commission (CCIH) with advisory and executive members who actively search for cases of hospital infection.

3.3 Sampling

The sample consisted of all patients who had a hospital-acquired infection admitted to the intensive care units (I and II) of the hospital with a positive culture result and/or clinical data suggestive of hospital-acquired infection according to the institution's protocol between June and September 2014.

3.4 Data Collection Instrument

The instrument used for data collection was a structured form that allowed data to be collected directly from the medical records of patients admitted to the intensive care units (I and II) with a diagnosis of hospital-acquired infection between June and September 2014.

The structured form is an effective tool for egalitarian research, the principle of which is to collect data resulting entirely from the researcher, i.e. used to indicate a number of questions that are investigated (MARCONI; LAKATOS, 2010).

3.5 Analysing data

The data was collected using a structured form, which was tabulated using the Microsoft Excel programme to calculate percentages and averages for analysis and comparison with the specific literature on the subject.

3.6 Ethical and Legal Aspects

The research carried out is based on Ordinance No. 422/2012, which guides the ethical and legal aspects of scientific research that guarantees the anonymity and confidentiality of participants, and also according to the recommendations of the National Health Council, through its Ordinance No. 466/2012 - Guidelines and Regulatory Standards for Research Involving Human Beings.

The research was registered on the Brazil Platform and sent to the Research Ethics Committee (CEP), and authorisation was also requested from the hospital to carry out the research. Data collection began only after the CEP had issued the research authorisation, authorised the institution's patient records service coordinator to carry out research on patient records (Appendix B) and agreed not to use the hospital's Free and Informed Consent Form (Appendix C).

The benefits of this research are the identification of microorganisms that are commonly **found in ICUs and are responsible for high** rates of hospital-acquired infections. By drawing up a profile of these microorganisms that cause HI, we can improve the quality of care provided to these patients, giving them a higher survival rate, especially in **ICUs.**

This study was carried out in two intensive care units (ICU) called ICU-1 and ICU-2, from June to September 2014, with a total sample of 33 patients with HI, 20 positive culture results, 11 types of isolated microorganisms were identified, the topography of these infections and the main antibiotics used by the patients evaluated, according to the information presented below from both ICUs.

Table 1: Socio-demographic profile of patients admitted with hospital-acquired infections to the ICUs of a public hospital between June and September. Teresina-PI, 2014.

Sex	n	%
Male	19	57,6
Female	14	42,4
Age group		
20 a 39	3	9,1
40 a 59	15	45,5
Over 60	15	45,5
Marital status		
Married	16	48,5
Single	14	42,4
Widowed	3	9,1
Origin		
Teresina	15	45,5
Interior of Piauí	14	42,4
Other states	4	12,6
TOTAL	33	100

Source: HGV ICU patient records

The research reveals that in relation to the gender variable, the distribution of male patients was the most frequent gender for ICU admissions with 57.6% (n=19) and females with 42.4% (n=14), it was also verified that the range of age groups varied between 20 and over 60 years, but the samples of patients with IH in the two ICUs between the ranges of 40 to 59 years and over 60 were n= 15 (45.5%) in both ICUs as shown in Table 1 presented.

It also shows that n=16 (48.5%) of the patients are married and 42.4% are single, with a further 9.1% who are no longer in any kind of relationship. The majority of patients came from the capital Teresina (45.5 per cent), followed by the interior of Piauí (42.4 per cent) and 12.6 per cent from other states.

In the study carried out on the 33 medical records surveyed, all of which showed clinical signs and a diagnosis of HI, only 13 were found to have positive culture results for HI, and the table below shows the topography of these infections.

Table 2: Distribution of infection episodes according to topography in ICUs 1 and 2, from June to September. Teresina-PI, 2014.

Topography	n	%
Respiratory	8	33,3
Operative Wound	7	29,2
Urinary	5	20,8
Blood	4	16,7
TOTAL	24	100

Source: Records of patients admitted to the HGV ICUs.

According to the results obtained through the survey, the distribution of infection cases by topography of patients admitted to ICUs 1 and 2, Table 2 shows the superiority of infections related to the respiratory system with 33.3% (n=8) of cases. This was followed by surgical wounds as the second most prevalent type of infection with 29.2% (n=7) of cases. The topographies with the lowest prevalence of infection were urinary tract (n=5) with 20.8% and bloodstream (n=4) with 16.7%.

Table 3: Distribution of microorganisms causing *hospital-acquired* infections in **ICUs - and 2, from June to September. Teresina-PI, 2014.**

Microorganism	n	%
Klebsiella spp	6	22,2
Acinetobacter spp	5	18,5
Pseudomonas aeruginosa	4	14,8

Escherichia coli	3	11,1
Staphylococcus aureus	2	7,4
Enterococcus	2	7,4
Stenotrophomonas maltophilia	1	3,7
Enterobacter spp	1	3,7
Proteus mirrabillis	1	3,7
Serratia marcenscens	1	3,7
Beta haemolytic Streptococcus spp	1	3,7
TOTAL	27	100

Source: HGV ICU patient records

Table 3 shows the predominance of HI by type of microorganism **isolated in the two ICUs of the hospital studied. This table shows Klebsiella spp** as the microorganism causing the highest number of infections with n=6, 22.2% of cases. **Acinetobacter spp** was the second most prevalent microorganism with n=5, 18.5%, followed by **Pseudomonas aeruginosa** n=4, 14.8%, and **Escherichia coli** with n=3, 11.1% of the cases. The microorganisms **Staphylococcus aureus** and **Enterococcus spp, had** equal prevalence of n=2, 7.4 per cent, and the other five microorganisms had lower prevalence of n=1 with only 3.7 per cent.

Looking at the first graph below, it can be seen that many of the hospitalised patients had delayed bladder catheterisation (DBS) (n=29) with 87.9% as the predominant invasive procedure. This was followed by the implantation of a central venous catheter (n=23) in 69.7% of cases. The predominant use of nasogastric tubes (NGS) was followed by the use of orotracheal tubes (OTT), both with a prevalence of 66.6% (n=22) of cases. The graph also shows that tracheostomy (TQT) was used in 30.3% (n=10) of patients, followed by other methods such as peripheral venous access (PVA) (n=9) with 27.3%, nasoenteral tube (NES) with n=7 patients (21.2%), and other procedures n=4 (12.1%).

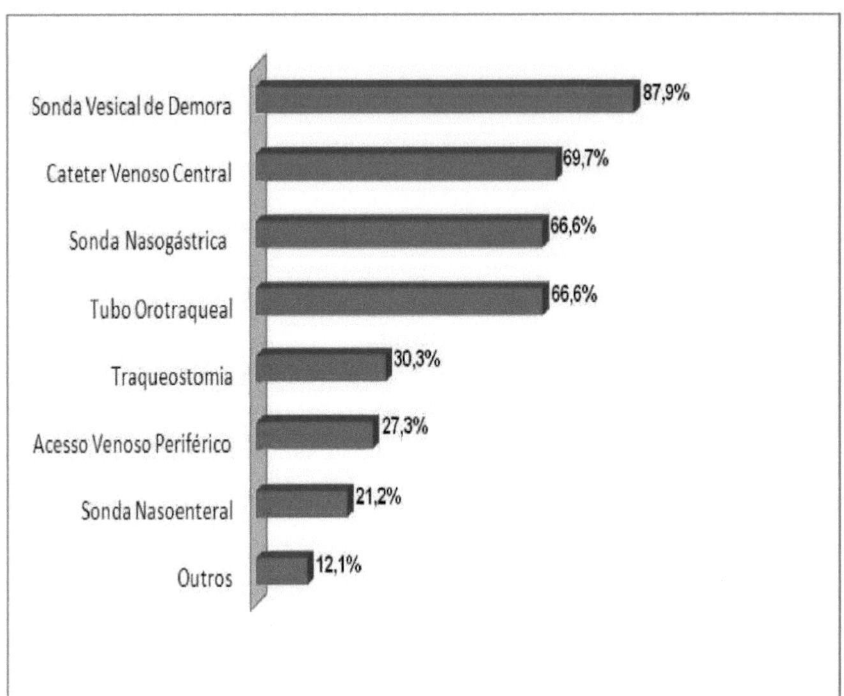

Graph 1: Distribution of types of invasive procedures performed in ICUs 1 and 2, between June and September. Teresina-PI, 2014.

Source: Records of patients admitted to the HGV ICUs.

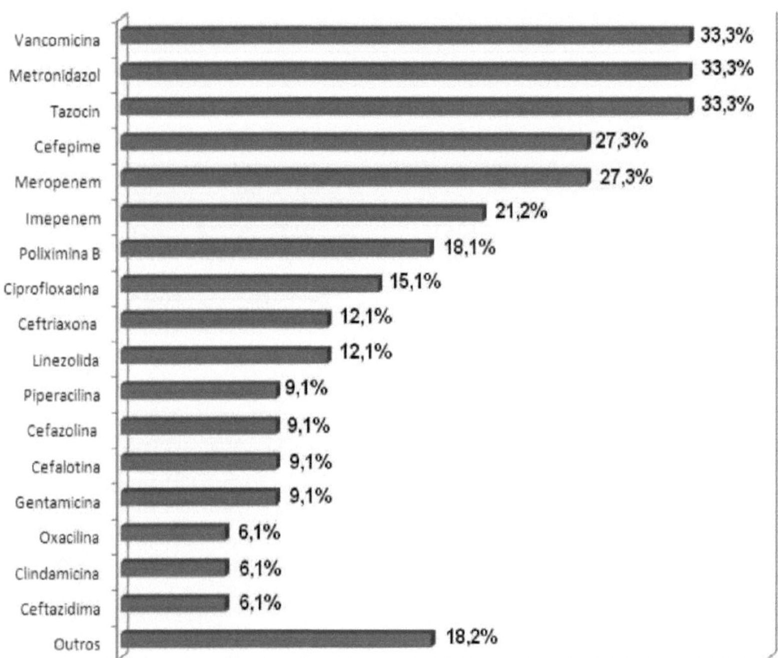

Graph 2: Distribution of the most commonly used antimicrobials in patients with hospital-acquired infections in ICU 1 and ICU 2, from June to September. Teresina-Pi, 2014.

Source: HGV ICU patient records

With regard to the most commonly used antimicrobials (Graph 2), Vancomycin, Metronidazole and Tazocin were the most prevalent, used in 33.3% (n=11) of the patients hospitalised with HI, and 17 different types of antibiotics were found in the two ICUs. Cefepime (n=9) and Meropenem (n=9) were the second most commonly used antibiotics, present in 27.3 per cent of cases, while several other antibiotics were also used, such as Imipenem (n=7) with 21.2 per cent, Poliximine B (n=6) with 18.1 per cent, Ciprofloxacin (n=5), Ceftriaxone (n=4), Linezolid (n=4), Piperacillin, Cefazolin, Cefalotin and Gentamicin (n=3) with the same prevalence of 9.1% of cases, as well as Oxacillin, Clindamycin and Ceftazidime (n=2) with a prevalence of 6.1%.

Among the other antimicrobials used were Piperacillin+Tazobactam, Amikacin, Tigecycline, Poliximine, Ampicillin and Ampicillin+Sulbactam, which were discarded after only one day's use by hospitalised patients.

Table 4: Antimicrobial sensitivity profile of microorganisms isolated in ICUs 1 and 2 between June and September. Teresina-Pi, 2014.

Sensitivity	Microorganisms			
	Klebsiela SPP	*Acinetobacter SPP*	*Pseudomonas SPP*	*Escherichia coli*
	n(%)			
Amikacin	6(100)	-	4(100)	3(100)
Ertapenem	6(100)	-	-	1(33,3)
Impenem	6(100)	-	1(25)	3(100)
Meropenem	6(100)	-	2(50)	-
Tigecycline	6(100)	5(100)	-	3(100)
Cefepime	3(50)	-	2(50)	3(100)
Cefotaxime	3(50)	-	-	1(33,3)
Ceftazidime	2(33,3)	-	1(25)	1(33,3)
Ceftriaxone	3(50)	-	-	1(33,3)
Gentamicin	4(66,7)	-	1(25)	3(100)
Ciprofloxacin	2(33,3)	-	3(75)	-
Piperacillin/tazobactan	-	-	2(50)	-
Levofloxacin	-	-	3(75)	-
Poliximine B	-	5(100)	3(75)	-
Ampicillin/Sulbactam	-	1(20)	-	-
Tobramycin	-	-	1(25)	-
Aztreonam	-	-	1(25)	-

Source: Records of patients admitted to the HGV ICU with positive culture results.

Table 4 shows the predominance of sensitivity of the microorganisms, considering that the four highlighted in the table were the most prevalent among the other microorganisms isolated. **Klebsiella spp** showed greater sensitivity to the carbapenem antibacterials Imipenem (n=6), Meropenem (n=6), Ertapenem (n=6), with 100% sensitivity, as well as to Amikacin (n=6).

Acinetobacter spp was sensitive only to Poliximine B and Tigecycline, both with (n=5)100%, and only (n=1)20% sensitive to Ampicillin/sulbactam. **Pseudomonas spp** was sensitive to Amikacin in 100% of cases, followed by the antimicrobials Ciprofloxacin, Poliximine B and Levofloxacin (n=3) with 75%. **Escherichia coli** (n=3) was 100 per cent sensitive to five types of antibiotic:

Amikacin, Imipenem, Tigecycline, Cefepime and Gentamicin.

Table 5: Discharge and mortality rates due to hospital-acquired infections for patients admitted to the ICUs of a public hospital between June and September. Teresina-PI, 2014.

Occurrence	n	%
High	14	42,4
Death	19	57,6

Source: HGV ICU patient records

With regard to the rate of discharge and mortality due to infection in ICU patients, death occurred in 57.6% of the sample obtained through the survey in both ICUs, followed by only 42.4% of patients who were discharged, as shown in Table 5.

CHAPTER 5

DISCUSSION

Intensive care units are particularly important for providing two main services to critically ill patients: life support for serious organ failure and intensive monitoring capable of enabling early identification and the necessary treatment of serious clinical complications, acting decisively when there is instability of organs and functional systems at risk of death (MOURA, et al, 2007).

Hospital infection rates in intensive care units vary between 18 and 54 per cent, and are around five to ten times higher than in other hospital inpatient units. The ICU is responsible for between 5 and 35 per cent of all HI and for approximately 90 per cent of all outbreaks that occur in these units, with high mortality rates ranging from 9 to 38 per cent, which can reach 60 per cent mainly due to the occurrence of HI (OLIVEIRA; et al, 2010).

Table 1 shows the sociodemographic profile of the patients in both ICUs. It highlights the predominance of males (n=19), with 56.6%, and also the prevalence of patients admitted as adults, aged between 40 and over 60 (n=15), with 45.5% of the cases,5% of cases, mostly patients in a stable relationship 48.5%, as shown in the study by Moura, et al (2011) also carried out in the intensive care units of the same hospital, where 56% of patients were male, 34.2% of them between the ages of 41 and 60, corroborating the findings of our research.

In the retrospective study by Abegg and Silva (2011) carried out between 2007 and 2008, with a total of 37 people diagnosed with an infection in the ICU in the city of Toledo in Paraná, 73% were male and the average age was 71. This was also found in Favarin and Camponogara's (2012) research into the profile of patients admitted to the ICU, which revealed that 42% of the patients were male, with the highest age range being 61 to 70 years.

In all the studies similar to the one carried out in the ICUs at the Getulio Vargas

Hospital (HGV), with regard to demographic characteristics, male patients were more likely to be hospitalised, which confirms the reality of healthcare systems in terms of the greater number of men who are more likely to receive this type of healthcare support than women.

With regard to the presence of patients aged between 41 and over 60, it can be inferred that the adult and elderly population requires a greater demand, which must be related to population ageing. We know that this group of people uses hospital services more intensively than other age groups, also causing an increase in cost, treatment and slow recovery.

Moura, et al (2011) in their study showed that the majority of patients came from the interior of the state (53.9%), unlike the rate found in this study, which had a higher rate of inpatients from the capital (n=15 (45.5%)) but was close to the rate of patients from other municipalities (n=14 (42. 4%)),4%), considering that this hospital is a reference point for treating highly complex diseases, this excessive demand ends up making it difficult to organise health care, since these patients who need specialised care need to be referred to hospitals in the capital.

Table 3 shows the prevalence of infection by topography in both **ICUs. The** most frequent HI is respiratory (33.3%), the second, surgical wound (29.2%) followed by urinary (20.8%) and blood (16.7%), The research carried out by Moura, et al (2007), and Moura (2011) corroborates the study carried out, stating that 59.4% of HI cases have the respiratory system as their main topography, considering that both studies were carried out in the same hospital, where it was found that this type of infection is mainly related to patients who use mechanical ventilators.

In the research carried out for this study it should be taken into account that of the total sample studied n=33, only 13 of these patients were diagnosed with infection through presumed clinical data and a protocol created by the institution where the research was carried out, this protocol also includes the National Nosocomial Infection Surveillance System (NNISS) methodology, which are established criteria for identifying and diagnosing hospital-acquired infections, the other 20 diagnoses of HI were confirmed through culture with a positive result.

Twenty cases of HI were laboratory confirmed, with **Klebsiella spp** (22.2%) as the predominant microorganism, followed by **Acinetobacter spp** (18.5%), **Pseudomonas spp** (14.8%) and **Escherichia coli** (11.1%). In the research carried out by Patricio (2008) these microorganisms were also identified as the main causes of HI, as well as in the study by Oliveira, et al (2010) and Moura, et al (2007), who state that **Klebsiella** is a bacillus present in the gastrointestinal tract of colonised individuals and an important pathogen causing HI, **especially in critical care units such as ICUs.**

Data from several other studies point to **Klebsiella spp** as a microorganism of great concern, since it has several resistance mechanisms and can trigger and be involved in several serious pathologies, which can lead to the death of several patients.

K. *pneumoniae* was the pathogen classically described by Friedlander (Friedlander's bacillus) as the cause of community-acquired lobar pneumonia, particularly in people with chronic alcoholism. This pneumonia is characterised by a severe pyogenic infection and high mortality rates. As an agent of community infections, **K. pneumoniae** is a pathogen with the potential to cause UTIs, pneumonia, bacteraemia and focal suppurative infections, including liver abscess and its serious complications, such as meningitis and endophthalmitis (WHOLHEIM, 2009).

In the Klebsiella spp. genus, an important mechanism of resistance is provided by broad-spectrum betalactamases (ESBL). These enzymes are capable of hydrolysing a wide variety of penicillins and third generation cephalosporins, which were initially developed as drugs capable of overcoming the bacterial resistance conferred by common betalactamases (OLIVEIRA, et al, 2011).

Henes, et al, (2013), in their research carried out in the ICU of a hospital in the state of São Paulo, revealed that during the observation period, there were seven admissions to this unit, making up the total capacity of the sector. Of the hospitalised patients, five (71% of the sample) were infected with **Acinetobacter baumannii**. In the review study by Fagon and Chastre (2002), in which data was collected and 2,490 isolates of respiratory tract microorganisms from 1,689 cases of ventilator-associated pneumonia were assessed, **Acinetobacter** was the second most common gram-negative organism, responsible for 8% of cases.

This microorganism can cause infection in any organ and the most frequently affected site is the respiratory tract. It is currently among the microorganisms most commonly involved in the aetiology of nosocomial pneumonia, along with **S. aureus, P. aueruginosa** and enterobacteria (OLIVEIRA, 2007).

Oliveira et al (2012b), during their study carried out between 2009 and 2010 in an emergency room of a University Hospital, with beds for patients who need continuous monitoring, or are in clinical and/or haemodynamic instability, identified **Acinetobacter spp.** as the most prevalent microorganism implicated in cases of HI isolated (37%), these data corroborate those found in our research, also in this study **Pseudomonas spp.** appears as a microorganism responsible for 11.1% of cases of HI infection.

Pseudomonas aeruginosa is the main pathogen of the Gram-negative bacteria group, causing infections in patients with deficient antimicrobial defences, and is considered an important hospital pathogen. These microorganisms are capable of causing a series of symptoms in patients, such as fever, shock, oliguria, leucocytosis and leucopenia, disseminated intravascular coagulation and adult respiratory distress syndrome (JAWETZ; MELNICK; ADELBERG; 2009).

According to Ferarezze (2007) in his study on hospital infection by multi-resistant **Pseudomonas**, 14.7% of the total cases were related to **Pseudomonas aeruginosa** as an isolated microorganism, mainly associated with cases of pneumonia, septicaemia and urinary tract infection. Moura (2007) emphasises that its pathogenesis should be discussed in the context of an opportunistic infection, which is capable of causing specific defects to some of the body's defence mechanisms and can colonise various tissues.

Escherichia coli, in this study as well as in the study by Oliveira et al, was classified among the most common resistant microorganisms also found in cases of HI in an intensive care centre, with a prevalence of 7.8% of cases.

Wollheim (2009) describes **E.Coli** as the most isolated pathogen (80 to 85%) in community-acquired urinary tract infections (UTIs) to a lesser extent (50%) at hospital level associated with structural abnormalities of the urinary system, stating that bacteraemias, respiratory infections and neonatal meningitis are caused by **E.coli** observed in elderly patients and newborns, with diseases that compromise the immune response, in hospitalised or institutionalised patients, usually associated with the presence of a delayed bladder catheter.

Positive cultures were also found with seven different types of microorganisms isolated, **Enterococcus spp, Stenotrophomonas maltophilia, Enterobacter spp, Proteus mirrabillis, Serratia marcescens, beta haemolytic Streptococcus spp.**

Enterococcus spp. are widely dispersed in nature and can be found in water, soil, food, plants and animals. In humans they are present in the microbiota of the gastrointestinal tract (GIT) and genitourinary tract (GUT) and, less frequently, in the oral cavity, gallbladder and male urethra. They can act as opportunistic pathogens, causing outbreaks of hospital infections that are difficult to control and spreading epidemic clones, are naturally tolerant to penicillins and resistant to cephalosporins, clindamycin and aminoglycosides (NACHTIGALL, 2011; COUTO 2010).

Strenotrophomonas maltophilia, currently the only species belonging to the genus **Stenotrophomona,** is characterised as a gram-negative non-glucose fermenting bacillus (GNFG), aerobic, with ubiquitous distribution and low virulence, considered an uncommon agent of infections, but it has been highlighted as an emerging nosocomial pathogen in immunocompromised patients, Its spectrum of infections is varied, including bacteraemias, skin and soft tissue infections, endocarditis, urinary tract infections, meningitis, intra-abdominal infections, ophthalmological syndromes and sinusitis. (RODRIGUES, et al, 2011).

In general, aminoglycosides are not active against **S. maltophilia,** probably because of inactivating enzymes and alterations to the cell surface present in the microorganism. Trimethoprim-sulfamethoxazole, a bacteriostatic agent, is the treatment of choice for infections caused by **S. maltophilia,** Ticarcillin-clavulanate is the only beta-lactam/betalactamase inhibitor combination that has an effective action and can also be used in patients intolerant to trimethoprim-sulfatoxazole (COUTO, 2010). **Enterobacter spp**, like **E.coli** and **Klebsiella,** are non-fermenting gram-negative enteric microorganisms that are considered opportunistic pathogens (SOUSA, et al, 2014).

Oliveira et al (2010), in their study, showed that the majority of patients used invasive procedures, including a delayed bladder catheter (68.5%) and a central venous catheter (49.6%), which also corroborates the procedures carried out on the patients studied in this study, as well as the use of nasogastric tubes and mechanical ventilation, both in 66.6% of cases. As identified in our research, studies have shown that the use of invasive procedures increases the risk of developing an infection.

Graph 2 shows that Vancomycin, Metronidazole and Tazocin are the most prevalent antimicrobials used by patients in both ICUs. Moura et al (2007) states that vancomycin is a glycopeptide currently in clinical use that is restricted to the treatment of gram-positive infections. The hospital where the research was carried out has a quick reference guide for health professionals on the anti-infectives available, for prescribing and rational use of this type of medication. Created in 2013, it has emerged as a tool for quick and easy access to the information needed so that professionals can prescribe with greater benefits and fewer risks.

Table 5 shows the sensitivity of **Klebsiella spp** to six different types of antimicrobial, amikacin, ertapenem, imipenem, meropenem, tigecycline all with 100 % sensitivity as well as **Acinetobacter spp** which proved sensitive to only two types of antibiotics, Tigecycline and Poliximine B, in the study by Nogueira (2009) carried out in a university hospital in Rio de Janeiro this microorganism proved sensitive only to poliximine B.

According to Couto (2010), this microorganism is an important cause of hospital-acquired infections, surpassed only by **pseudomonas aeruginosa,** and has been implicated in several HAs. It generally has numerous mechanisms of resistance to antimicrobials, and is related to resistance to beta-lactam antibiotics and the production of beta-lactamase.

Pseudomonas aeruginosa is a ubiquitous microorganism that has the ability to grow in conditions of low nutrient availability and temperature extremes. It is intrinsically resistant to various types of antibiotics, with the presence of a combination of resistance mechanisms that give it the characteristic of antibiotic multidrug resistance, It proved to be 100 per cent sensitive to amikacin, just as other studies have revealed sensitivity (64.9 per cent) of this microorganism to this type of antibiotic, and 62.2 per cent to meropenem, unlike this study which only revealed sensitivity in 50 per cent (n=2) of cases of HI caused by this pathogen (COUTO, 2010).

E. coli, according to Couto (2010) is one of the most common gram-negative bacilli associated with HI and the first among the agents of urinary tract infections, resistance to broad-spectrum cephalosporins is basically mediated by ESBL. Most ESBLs have increased activity against ceftazidime and aztreonam, and decreased activity against cefotaxime, although the opposite may be true in some cases. In this study, **E.coli** proved to be 100 per cent sensitive to Amikacin, Impenem, Tigecycline, Cefepime and Gentamicin.

In general, ESBLs are enzymes that hydrolyse penicillins, first-, second- and fourth-generation cephalosporins and the monobactam aztreonam. However, they do not act on cefamines and carbapenems. In contrast, ESBLs are inhibited by commercially available beta-lactamase inhibitors, including clavulanic acid, tazobactam and sulbactam.

The HI rates were (n=3) 3%, (n=18) 20%, (n=11) 13% and (n=1) 1% respectively for the months of June, July, August and September, which were the months studied. Table 5 shows the mortality and discharge rates of the patients in the study. The death rate was higher (57.6%) than the rate of patients discharged (42.4%), considering that the patients discharged from the ICU continued to be hospitalised.

Among the patients studied by Oliveira et al (2010), of the 195 deaths (10.3%) that occurred, 39.5% (77) were from patients who developed HI, as well as in the retrospective study by Guimarães et al (2011) where deaths from hospital-acquired infections totalled n=77, 56.4% of cases and around 110 (82.7%) related to multidrug-resistant microorganisms; these findings corroborate our research.

According to Cavaleiro (2011), hospital-acquired infections account for around

50 per cent of the total mortality caused by healthcare-associated infections. It is therefore essential for staff to keep up to date with the resident flora and its resistance pattern in the hospital unit where they work, since this differs between units and changes over time, as the acquisition of resistance is a dynamic process.

CHAPTER 6

CONCLUSION

According to the findings of our research, and relating them to our objectives, it was observed that HIs affecting the respiratory tract are the most prevalent among others within ICUs, **Klebsiella spp** stood out as the most prevalent isolated microorganism causing HIs in the hospital under study, accounting for 22.2 per cent of infection cases.

Among the most commonly used antimicrobials, Vancomycin, Metronidazole and Tazocin were used about eleven times (33.3%) in cases of HI. As for the sensitivity of the microorganisms to the antibiotics used, **Klebsiella spp.** was more sensitive to the carbapenem class (ertapenem, imipenem, meropenem), Amikacin and Tigecycline, the latter of which was also effective (100%) against the coccobacillus **Acinetobacter,** accompanied by Poliximine B. **Pseudomonas aeruginosa** was only 100 per cent sensitive to amikacin, followed by (75 per cent) ciprofloxacin, levofloxacin and poliximine B, and 50 per cent to the antimicrobials meropenem, cefepime and piperacillin+tazobactam. **E.coli** showed 100 per cent sensitivity to the following antibiotics: amikacin, imipenem, tigecycline, cefepime and gentamicin.

Based on the findings of this study, it is suggested that future studies be carried out, also through periodic discussions between the health teams in this same hospital, with the aim of analysing these indicators of hospital infection rates, microbial resistance profile, in the perspective that these IH could be avoided, working together with the institution's CCIH, promoting guidelines and continuing education activities, encouraging the participation of teams to increase infection control measures, and providing information for the creation of antibiotic prescription protocols according to the resistance profile of microorganisms and that a periodic review of these protocols can also be carried out.

The data obtained through this study show us how necessary it is to study these infections in intensive care units, given the emergence of these microorganisms which, when associated with an underlying pathology and the vulnerabilities of patients admitted to intensive care units, cause a large number of patient deaths.

In addition to the healthcare team's knowledge of these microorganisms, it is extremely necessary to discuss and carry out continuing education activities related to infections caused by MR, including care practices, control and prevention of these infections through guidance from nursing management and especially with regard to hand washing, which is one of the main ways of reducing the risk of transmission of infection in hospitals and especially in ICUs, nursing management must be attentive to the care and manner in which the team deals with these patients, even with family members who frequently visit them and have no guidance on hand washing and its importance both for themselves and for their hospitalised family member.

Together with nursing management and the ICHR, it is of fundamental importance for these professionals to have education and guidance on the importance of preventing the emergence of cases of infection, knowing that in addition to causing high hospital costs, when associated with resistant microorganisms it can further increase the morbidity and mortality rate within these units.

CHAPTER 7

REFERENCES

ABEGG, P.T.G.M., SILVA, L.L. Hospital infection control in an intensive care unit: a retrospective study. Semina: Ciências Biológicas e da Saúde, Londrina, v. 32, n. 1, p. 47-58, jan./jun. 2011.

ALMEIDA, A.A., MANFRÉ L.L.M., CRODA, M.T.R.C.G., CHANG, M.R, OLIVEIRA, K.M.P. Factors associated with klebsiella spp. bacteraemia in a university hospital. Rev. Evidência, Joaçaba. v.12.n.2.p.165-174. julh/dez. 2012.

BRAZIL. Ministry of Health. Ordinance N° 2.616, of 12 May 1998. Hospital Infection Control Programme. Official Gazette, Federative Republic of Brazil, Brasília (DF), Jul 1998.

CAVALEIRO, P.L.G. Prevention of Nosocomial Infection in Intensive Care Units. Dissertation (Master of Medicine) - University of Porto, Porto, Portugal. 2011.

CHASTRE, J. FAGON JY. Ventilator-associated pneumonia. Am J Respir Crit Care Med. 165:867-903. 2002

COUTO, C.R., PEDROSA, G.M.T., CUNHA, A.F.A., AMARAL, B.D. Infecção Hospitalar e outras complicações não-infectiosas da doença. 4ª Ed. Guanabara koogan, 2010.

FAVARIN, S.S., CAMPONOGARA, S. Profile of patients admitted to the adult intensive care unit of a university hospital. Rev. Enferm. UFSM May-Aug; 2(2): 230-329. 2012.

FREIRE I.L.S., ARAUJO., R.O., VASCONCELOS, Q.L.D.A.Q., MENEZES, L. C. C., COSTA, I.K.F., TORRES, G. V. Microbiological, sensitivity and bacterial resistance profile of blood cultures from a paediatric intensive care unit. Rev. Enferm. UFSM. Sep/Dec. 2013.

FERRAREZE M.V.G., LEOPOLDO V.C., ANDRADE D., SILVA M.F.I., HAAS V.J. Multiresistant Pseudomonas aeruginosa in intensive care units: challenges ahead? Rev. Acta Paul Enferm.,2007.

GIL. A C. Como elaborar projetos de pesquisa. São Paulo: Atlas. 5th ed. 2010

GUIMARÃES A C.. DONALISIO M R., SANTIAGO T.H.R. FREIRE, J.B. Deaths associated with hospital-acquired infections in a general hospital in Sumaré, SP, Brazil. Rev. Bras. Enferm, Brasília Sep-Oct;64(5):864-9. 2011.

HENES MA, SILVA SC, FORNARI JV, BARNABÉ AS, FERRAZ RRN. Incidence of acinetobacter infection in a special treatment unit of a public hospital in the state of São Paulo .Science in Health - May-Aug; 4(2): 9710. 2013.

JAWETZ; MELNICK; ADELBERG; Medical microbiology: a Lange medical textbook.
24. ed. Rio de Janeiro: McGraw-Hill interamericana do Brasil LTDA. 2009.

MOURA M.E.B., CAMPELO S.M.A., BRITO F.C.P., BATISTA O.M.A., ARAÚJO T.M.E., OLIVEIRA, A.D.S. Infecção Hospitalar: estudo de prevalência em um hospitalo publico de ensino. Rev. Bras Enfermagem, vol 60,n.4,p.416-421,2007.

MOURA M. E.B., NUNES M. R. C. M., ARAUJO. TELMA M. E. A. MONTEIRO, C. F. S., CARVALHO, L. R. B. Hospital infections in intensive care units in a public hospital. Rev. Interdisciplinar NOVAFAPI, Teresina. V.4, n.4, p. 42-48, Oct-Nov-Dec. 2011.

MARCONI, A.M., LAKATOS, M. E.. Research techniques.7ª Ed. Editora Atlas, 2010.

NACHTIGALL, G. Evaluation of the diversity and susceptibility profile of Enterococcus sp. Isolated in the waters of Arroio Dilúvio. - (Master's thesis Federal University of Rio Grande do Sul, Institute of Basic Health Sciences, Postgraduate Programme in Agricultural and Environmental Microbiology) Porto Alegre, RS-BR. 2011

NOGUEIRA, P.S.F., MOURA, E.R.F., COSTA, M.M.F., MONTEIRO, W.M.S., BRONDI L. Profile of hospital infection in a university hospital. Rev. Enferm. UERJ, Rio de Janeiro, Jan/Mar; 17(1): 96-101. 2009

OLIVEIRA, C.B.S, DANTAS., V,C.B, NETO R M., AZEVEDO P.R.M, MELO M.C.N.M. Frequency and resistance profile of Klebsiella spp. in a university hospital in Natal/RN during 10 years . J Bras Patol Med Lab - v. 47 - n. 6 - p. 589-594. december 2011

OLIVEIRA, A.C., KOVNER, C.T., SILVA, R.S. Hospital infection in an intensive care unit of a Brazilian university hospital. Rev. Latino- Americana de Enfermagem, v.18, n.2,p.98-104, 2010.

OLIVEIRA, A.C., ANDRADE, F.S., DIAZ, M.E.P., IQUIAPAZA, R.A. Colonisation by microorganisms and infection related to health care. Acta Paul enferm. 2012; 25(2):183-9. (b)

OLVEIRA, M,S.Treatment of infections caused by carbapenem-resistant Acinetobacter spp. Dissertation presented to the São Paulo Medical School. 2007

OLIVEIRA A.C, PAULA A.O, IQUIAPAZA R.A, LACERDA A.C.S. Healthcare-related infections and clinical severity in an intensive care unit. Rev Gaúcha Enferm. 2012;33(3):89-96. (a)

OLIVEIRA A.C., GONZAGA C. R., DAMASCENO Q.S., GARBACCIO,J.L. Challenges and perspectives for the containment of bacterial resistance from the perspective of health professionals. Rev. Eletr. Enf. [Internet]. 2013 Jul/Sep.

PRADE, SS. Brazilian Study of the Magnitude of Hospital Infections in Tertiary

Hospitals. Rev. Controle Infecção Hosp. 1995.

PATRICIO, M.I.A. Analysis of data on infections in intensive care units (ICU) of tertiary level hospitals in fortaleza, state of ceará, from january 2005 to december 2007. (Dissertation Professional Master's Degree in Health Surveillance) - Sergio Arouca National School of Public Health - ENSP, Fortaleza, Ceará.

RODRIGUES L.S., DI GIOIA T.S.R., ROSSI F. Stenotrophomonas maltophilia: emerging resistance to SMX-TMP in Brazilian isolates. A reality? Bras. Patol Med Lab, V.47, n.5. p. 511-517. Oct-2011.

SOUSA R.R.F., Search for quinolone resistance genes in gram-negative bacilli of clinical and environmental origin. Dissertation for the postgraduate programme in Public Health at the School of Public Health of the University of São Paulo. SP- BR, 2014.

WOLHEIM C. Molecular epidemiology of extended-spectrum bata-lactamase-producing Escherichia coli and Klebsiella spp. PhD Thesis - Postgraduate Programme in Biotechnology, University of Caxias do Sul. RS - BR. 2009.

CHAPTER 8

APPENDIX

APPENDIX A
SEMI-STRUCTURED FORM

Location: ICU-I ( ) ICU-II ( )   Date: _____ / _____ / ______

Record number: Enf/Leito:

Sex: M ( ) F ( ) Age: __________ Profession: ______________________

Marital status: ______________ Origin: ______________________

Medical Diagnosis: ______________________________________

Type of treatment: ( ) Clinical ( ) Surgical

Hosp admission:___/___/___ICU admission:___/___/___Discharge:___/___/___

Type of discharge: ______________________________________

____

____

____

Invasive procedures: ______________________________________

____

____

____

Have you used antibiotics: yes ( ) no( )

If yes, which antibiotic(s) were used and for how long:

1   Home /   /__ End _______ /___/____

2   Home /   /__ End _______ /___/____

3   Home /   /__ End _______ /___/____

4   Home /   /__ End _______ /___/____

5   Home / /___ End _________ / ___/____

6   Home / /___ End _________ / ___/____

7   Start / / End___ /_________ / ____

8   Home / / End___ /_________ / ____

9   Home / / End___ /_________ / ____

10   Start / / End___ /_________ / ____

CULTURE+ANTIBIOGRAM PERFORMED: YES( ) NO ( )

Topography site: ____

Date of collection:___/___/___Date of   result:___/___/___

Isolated microorganism: _______________________________________

____

____

____

Sensitive: ____

____

____

____

Resistant: ____

____

____

____

Intermediate: ____

____

____

____

APPENDIX B

AUTHORISATION FORM FOR RESEARCH ON MEDICAL RECORDS

Your Excellency the Coordinator of the Patient Record Service at Getúlio Vargas Hospital

I, Marcelo de Moura Carvalho, who is primarily responsible for the research of participants Anna Karoeny Da Silva Santos and Jackeline Alves de Araujo, which belongs to the Nursing course at the Associação de Ensino Superior do Piauí - AESPI, hereby request authorisation to collect data from the medical records of patients undergoing treatment in intensive care units between June and September 2014. This data will subsidise the work entitled Microbiological Profile of Hospital Infections in Intensive Care Units, which aims to evaluate the microbiological profile of hospital infections in the intensive care units of the aforementioned hospital.

Teresina, 5th September 2014.

APPENDIX C

CONSENT FORM FOR NON-USE OF THE INFORMED CONSENT FORM

Honourable Coordinator of the Research Ethics Committee of Universidade Paulista (UNIP):

Concerning the final year research project by participant(s) Anna Karoeny da Silva Santos and Jackeline Alves Araújo, which belongs to the nursing course, Associação de Ensino Superior do Piauí- AESPI, entitled Microbiological Profile of Hospital Infections in Intensive Care Units.

Researchers: Marcelo de Moura Carvalho - supervisor
Anna Karoeny da Silva Santos and Jackeline Alves de Araújo

We would like to make it clear that the data for this research will be collected using clinical records, and that it will not be possible to locate each of the subjects who are part of the research sample. We therefore request the express agreement of the Coordinator of this Committee not to use the Informed Consent Form.

Teresina- PI, 21st August 2014.

ANNEXES

ANNEX A

Vice-Reitoria de Pós-Graduação e Pesquisa

UNIVERSIDADE PAULISTA – UNIP
Campus Indianópolis
Rua Dr. Bacelar, 1212 - 4º andar – Vila Clementino
CEP: 04026-002 – F. (11) 5586-4090
E-mail: cep@unip.br

COMITÊ DE ÉTICA EM PESQUISA

CARTA DE ANUÊNCIA

Declaramos para os devidos fins, que aceitaremos a pesquisadora, Anna Karoeny da Silva Santos e Jackeline Alves de Araújo acadêmicas do curso de Enfermagem da Associação de Ensino Superior do Piauí - AESPI, a desenvolver o seu projeto de pesquisa, **PERFIL MICROBIOLÓGICO DAS INFECÇÕES HOSPITALARES NAS UNIDADES DE TERAPIA INTENSIVA DE UM HOSPITAL PÚBLICO** que está sob a orientação do Professor Msc. Marcelo de Moura Carvalho, cujo objetivo é identificar os principais microorganismos causadores de infecção hospitalar em duas unidades de Terapia Intensiva em um hospital público de referência em Teresina- Pi.

A aceitação está condicionada ao cumprimento dos requisitos da resolução 466/12 e suas complementares, comprometendo-se a utilizar os dados e materiais coletados, exclusivamente para os fins da pesquisa e coleta de dados após a aprovação do CEP.

Teresina – PI, em 9 / 7 / 14

Nome/ assinatura e carimbo do responsável pela Instituição

UNIP AUTHORISATION LETTER

UNIVERSIDADE PAULISTA - UNIP - VICE-REITORIA DE PESQUISA E PÓS	

PARECER CONSUBSTANCIADO DO CEP

DADOS DO PROJETO DE PESQUISA

Título da Pesquisa: PERFIL MICROBIOLÓGICO DAS INFECÇÕES HOSPITALARES NAS UNIDADES DE TERAPIA INTENSIVA EM UM HOSPITAL PÚBLICO

Pesquisador: MARCELO DE MOURA CARVALHO

Área Temática:

Versão: 1

CAAE: 35928214.0.0000.5512

Instituição Proponente: Universidade Paulista - UNIP / Vice-Reitoria de Pesquisa e Pós Graduação

Patrocinador Principal: Financiamento Próprio

DADOS DO PARECER

Número do Parecer: 826.402

Data da Relatoria: 09/10/2014

Apresentação do Projeto:

PROPOSTA SE ENCONTRA NAS NORMAS CIENTÍFICAS

Objetivo da Pesquisa:

- EXEQUÍVEIS

Avaliação dos Riscos e Benefícios:

- RISCO MÍNIMO

Comentários e Considerações sobre a Pesquisa:

- TEMA RELEVANTE PARA A SAÚDE

Considerações sobre os Termos de apresentação obrigatória:

- TERMOS OBRIGATÓRIOS APRESENTADOS

Endereço: Rua Dr. Barcelar,1212
Bairro: Vila Clementino   **CEP:** 04.026-002
UF: SP   **Município:** SAO PAULO
Telefone: (11)5586-4090   **Fax:** (11)5586-4073   **E-mail:** cep@unip.br

Continuação do Parecer: 826.402

Recomendações:
- SUGIRO ACRESCENTAR QUE OS SUJEITOS SERÃO MAIORES DE 18 ANOS PARA PODER UTILIZAR O TCLE.
- SUGIRO DIVULGAÇÃO DOS RESULTADOS AO FINAL DA PESQUISA

Conclusões ou Pendências e Lista de Inadequações:
- PROJETO SE ENCONTRA METODOLOGICAMENTE CORRETO E RESPEITA AS NORMAS CIENTÍFICAS E ÉTICAS

Situação do Parecer:
Aprovado

Necessita Apreciação da CONEP:
Não

Considerações Finais a critério do CEP:
- SUGIRO ACRESCENTAR QUE OS SUJEITOS SERÃO MAIORES DE 18 ANOS PARA PODER UTILIZAR O TCLE.

SAO PAULO, 09 de Outubro de 2014

Assinado por:
JOSE BARBOSA
(Coordenador)

Endereço: Rua Dr. Barcelar,1212
Bairro: Vila Clementino   CEP: 04.026-002
UF: SP   Município: SAO PAULO
Telefone: (11)5586-4090   Fax: (11)5586-4073   E-mail: cep@unip.br

ANNEX C
ABSTRACT STATEMENT

DECLARAÇÃO

EU, **ANTENOR DE SOUSA LIMA FILHO**, brasileiro, casado, com formação em Letras Inglês pela Universidade Estadual do Piauí – **UESPI**, Professor da Rede Pública e Particular de Teresina, portadordo RG: **821.732** - SSP/PI, CPF: **287237913-49**, residente e domiciliado na Quadra – 112 Casa - 14, Bairro Dirceu Arcoverde I, Teresina-PI, declaro para os devidos fins que prestei serviço de tradutor para aluno da Associação de Ensino Superior do Piauí -**AESPI**, Anna Karoeny da Silva Santos e Jackeline Alves Araújo, em trabalho de monografia para conclusão do curso de **ENFERMAGEM**, tendo como tema " **Perfil Microbiológico das Infecções Hospitalares nas Unidades de Terapia Intensiva**". Ratifico serem verdadeiras as informações acima prestadas.

Teresina, 04 de dezembro de 2014

Antenor de Sousa Lima Filho
Nome do Tradutor

I want morebooks!

Buy your books fast and straightforward online - at one of world's fastest growing online book stores! Environmentally sound due to Print-on-Demand technologies.

Buy your books online at
www.morebooks.shop

Kaufen Sie Ihre Bücher schnell und unkompliziert online – auf einer der am schnellsten wachsenden Buchhandelsplattformen weltweit! Dank Print-On-Demand umwelt- und ressourcenschonend produziert.

Bücher schneller online kaufen
www.morebooks.shop

Printed by Books on Demand GmbH, Norderstedt / Germany